STEM:

A CALL TO LEADERSHIP

Cover Artwork by Ella Beard
Design by Reyhana Ismail

All inquiries or sales requests should be addressed to:

PLANTING PEOPLE
GROWING JUSTICE

Planting People Growing Justice Press
P.O. Box 131894
Saint Paul, MN 55113
www.ppgjbooks.com

Printed and bound in the United States of America
First Edition
LCCN: 2023943251
SC ISBN: 978-1-959223-27-6

TABLE OF CONTENTS

About Planting People Growing Justice 4

From The Editor 5

Foreword 6

How Can Stem Change The World? 7

Stem And Education 13

Justice And Equity 17

Careers That Make A Difference 21

Stem: A Call To Leadership 25

Reflection Guide 29

ABOUT PLANTING PEOPLE GROWING JUSTICE LEADERSHIP INSTITUTE

Planting People Growing Justice Leadership Institute seeks to plant seeds of social change through education, training, and community outreach.

All proceeds from this book will support the educational programming of the Planting People Growing Justice Leadership Institute.

Learn more at www.ppgjli.org.

PLANTING PEOPLE GROWING JUSTICE

*This activity is made possible through the support of The Science Museum of Minnesota and The Donaldson Foundation.

Science Museum of Minnesota

SCIENCE IS ALL OF US

Donaldson. FILTRATION SOLUTIONS

LETTER FROM THE EDITOR

My STEM learning journey began at the Science Museum of Minnesota. I started working at the museum as a lab partner in the Experiment Gallery when I turned thirteen years old. This was my first job, yet it had a lasting, lifelong impact. Still today, I leverage STEM skills in my daily work. These skills provided me with a solid foundation for my future career as an educator and lawyer.

I developed the skills that I needed to succeed by:

- Learning the importance of working with teams as our Youth Science Center group developed the River Mall exhibit together.
- Engaging in research on current issues like renewable energy and race and science.
- Solving problems in innovative ways, whether it be working to end pollution or to protect an endangered species.
- Communicating new ideas, which led to the creation of an exhibit to honor pioneering African Americans in Science like Drs. Frank Snowden (chemical engineer) and Patricia Bath (ophthalmologist).

Today, I seek to share this gift of STEM education with our youth. I seized this opportunity to help them develop new skills and engage in career development. This is vitally important when STEM-based careers are growing at a rate of two times faster than all other careers combined (Bureau of Labor Statistics). STEM education is the preparation for the jobs of today and tomorrow.

STEM provides skills and tools that are essential for all careers by training the next generation of creatives, innovators, and disruptors.

DR. ARTIKA R. TYNER

Editor

Planting People Growing Justice Leadership Institute

FOREWORD

As leaders at the Science Museum of Minnesota and enthusiastic supporters of the work at the Planting People Growing Justice Leadership Institute, we are proud and honored to have had a role in the creation of this anthology. We hope you'll find it captures both the inspiring possibilities of STEM and the extraordinary promise of future world-changers.

The Science Museum and Planting People Growing Justice have a common goal—to empower young people to be innovative thinkers and visionary leaders who will work together to create a more just and equitable world.

We use multiple strategies to progress towards this goal, but the concept at the heart of our collective work is showing our young people that their perspectives, ideas, and voices are more important than ever. By encouraging them to reflect on the impact of STEM in the world around them, we are supporting them to lift their voices to solve the most pressing problems facing our communities.

At the Science Museum, science and education centering equity is our superpower. Led by our Statement on Equity and Inclusion, we are actively working to undo oppressive systems and use STEM as a tool to advocate for justice and equity. We envision a world where everyone has the power to use science to make lives better, and we are continually seeking out ways to provide learning opportunities that inspire discussion, reflection, and active participation in shaping our communities.

Alongside organizations like PPGJLI, who we are joining in doing this vital work, we have great hope for a bright future. The STEM learners of today are the STEM leaders of tomorrow, and together, we are committed to showing them that there is a place for them in science. We hope you enjoy this window into the minds of our next generation as much as we have.

ALISON REMPEL BROWN

President and CEO

Science Museum of Minnesota

JOANNE JONES-RIZZI

Vice President of Science, Equity, and Education

Science Museum of Minnesota

HOW CAN STEM CHANGE THE WORLD?

"I know of nothing more inspiring than that of making discoveries for one's self."

Dr. George Washington Carver

HOW CAN STEM CHANGE THE WORLD?

SKYE BISHOP

How can STEM be used to solve a problem in your community? STEM can be used in a lot of different ways and scenarios. We could break STEM down into its individual letters. Let's use a scenario.

You are trying to host a party. You need to know what kind of food, activities, and decorations to use. Also, how to create and buy these items. This scenario is only an example and is not a problem that needs to be solved emergently.

S: Science. Science is a good first step in solving problems in the real world. Science is also a very general term. It is a subject that is broken down into many parts. Biology, geology, meteorology, etc., although in most situations, we can use science as a general term. In the case of a real-life scenario, we could use the scientific method. The scientific method is a step-by-step procedure. First, define a question. Next, create a hypothesis. After that, research and gather data on your topic. Then, look at your data. Finally, draw a conclusion. In our scenario, we are going to first, define our question of what food, activities, and decorations do attendees want? Second, we'll come up with a couple of options for each category. Third, we need to poll the people who are coming to the party. Fourth, we will compare all the results from our polls. Fifth, we're going to decide on the most popular food, activities, and decorations.

T: Technology. Technology can be used in many different ways. Technology is the process of things working, although it is also considered just electronics. We can use electronics to organize the results. After we poll our attendees, we can organize them in an app on our phone or computer. We could also use technology to do research on our topic. We could use our computer to look at certain areas or to ask people about subjects from across the world.

E: Engineering. Engineering is very useful. It is involved in almost every world issue. For now, we are going to stick to our scenario. We can use engineering to create decorations and activities. Engineering can help us make sure the materials we use in our projects will look good together. It can also insure that they will last through the party and beyond. Engineering is important in our real-world scenarios. It can be helpful in creating inventions that will help in everyday life.

M: Mathematics. Math is intertwined with all these other areas in STEM. It is almost directly involved in science. Math gives a factual reasoning behind science. Math also helps in creating technology. Engineering is supported by math so that buildings and inventions work. Without math, science, technology, and engineering may cease to exist. In our scenario, math will help us compare numbers from the poll. It will support our technology by making sure it makes logical sense. Then, the math will get the engineering started and working right.

Hopefully this opened your eyes to the wonders of science, technology, engineering, and mathematics. These subjects are constantly in your life—every second of it—and you may not even realize it.

Thank you for reading my small essay. Have a great day.

Sincerely,

Skye Bishop.

STEM CAN

FOREVER MASON

STEM can:

Make people smarter by discovering something new.

Make people happier by discovering things like ice cream.

Change a person's reality by finding something they want to do and never had a chance to try.

STEM CAN CHANGE THE WORLD

AMAL FARAH

STEM will help our environment grow. It will help us people join together to make the next big thing in our blue-green world. STEM will empower kids like me from around the world to be more confident in themselves. But not only that—our ecosystems will be able to be inhabited by more animals. And we'll find a way to reduce fossil fuels. That's how STEM will change the world.

LAMAR

STEM can change the world by using science to see what can be made to make pollution better. Technology can be used to collect pounds of trash from the ocean and make a safer environment.

MASON

Science can help people that are sick.

Technology right now helps us by building stuff.

Engineering uses math and science skills by designing stuff for use.

Mathematicians are skilled, solve problems, and help us to make our brains grow.

STEM'S SUPERPOWER

AYA DAALI

STEM is full of power. It's like your own superpower.

If you know it, it is divine. And it whirls around in your mind.

From sky to space to underwater lakes.

Growing technology and advancing growing minds.

For those who care what STEM provides.

Some might say that it is dumb, but STEM is the base of what they may become.

S is for Science.

T is for Tech.

E is for Engineering, you might expect.

M is for Math, the base of it all.

But be careful, as the embrace might look small.

It's everywhere, even at the mall.

From TVs to speakers... to your oven heater.

STEM can change the world by... stopping global warming and making the world thrive.

STEM AND EDUCATION

"I like to learn. That's an art and a science."

Katherine Johnson, Mathematician

STEM, THE POEM

AYA DAALI

In the summer when there is no school,
But very close, say goodbye to the pool.
Students and parents swarm the store
In hopes they will get supplies before there is no more.

But meanwhile in an office,
The principle is hiring teachers that will be flawless.
But someone comes in unrecognized,
Someone with science in their eyes.
That someone had a thing in mind.
That thing was to bring STEM into students' minds, so they could shine!
The principal hesitated.
But after some explanation, he was fascinated.
He thought that was a brilliant idea.
He hired her on the spot.
Her name was Mia.

When you get back to school,
Where you will learn not to be a fool,
You are shocked at the sight.
That the once-abandoned room had some light.
You peek inside in wonder.
Then, you realize you had made a blunder.
Someone is in that room,
Teaching a class so you let them resume.
When it is finally your time to go
Into a room that you don't know.
You want to ask so many questions,
But Ms. Mia starts a lesson.

She says that STEM is more than just a word.
It is science, technology, engineering, and math.
You heard.
In this class you learn many things,
Like how STEM changes the world.
By using science to help the dying world.
She tells you about a time before solar panels.
The world was full of CO_2 (Carbon Dioxide), it was about to dismantle.
But one man (Charles Fritts) came up with an idea.
"To use the sun to make energy," said Ms. Mia.

We use STEM in many different ways.
To upgrade technology and make new games.
An alert goes on the news.
That someone needs funding for what he calls "Project Rocket Fuel."
But no one wants to donate to him.
You had no idea how to help, but the idea came on a whim.
"So we should protest," you said to your class.
Your teacher agreed, this idea would surpass!

You wrote on posters in your way:
Donate to STEM, it's the reason the world is what it is today.
After the protest, to your delight,
Rocket shoes were announced that night.
You are glad, because of STEM...
That people could see the science in them.
You hope that STEM should keep going.
Because who knows what possibilities the world will be holding.

Come on! You can do STEM too!
By doing anything, like asking a question: what, where, why, when, and who?
As long as STEM stays alive,
We will live and a happy world will thrive.

BENJAMIN THOMAS

STEM is important for education and solving worldwide problems.

Pathologists study diseases and help to make cures for animals and people who are sick.

Engineers design sturdy structures and design machinery.

TORI SELLNER

STEM is

Creative Innovation

STEM CAN CHANGE THE WORLD BY

ARI' SIMONE

Having more hands-on things for people to learn so other people can also learn and adapt.

JUSTICE AND EQUITY

"Every scientist dreams of doing something that can help the world."

Tu Youyou, Scientist and Nobel Prize Laureate

GIRLS IN SCIENCE AND TECHNOLOGY POEM

HAWA FARAH

Science and Technology can and will open new doors for girls.

STEM improves our minds and daily lives.

It will help our minds think logically and critically.

STEM inspires us girls to dream big and bold.

It gives us opportunities to lead and advocate and empower others.

STEM improves our minds and daily lives.

Increase the amount of service and labor instantly.

Science and Technology expand human communication around the globe (future TikTok, Snapchat, Instagram, etc.).

It will solve environmental problems, such as reducing fossil fuels.

STEM improves our minds and daily lives.

Ecosystems will be safer with more renewable resources from science and technology experts.

STEM will improve girls like myself to be more creative and imaginative.

STEM will enhance mobility and future ways to preserve fossil fuels.

STEM will improve others' minds and daily lives.

Girls like me are the heart of the generation.

And using stem, girls like me will create the next BIG thing.

STEM CAN CHANGE THE WORLD BY

CORDELL

Advancing technology so humans would not have to work and money would not be a thing.

ROLAND HESTER

STEM can change the world because technology and jobs can help you get smart and get paid, too.

It can help you get good at what you are struggling with.

It is good to use and can help with more than just technology and help your writing and reading, too.

HOW STEM COULD CHANGE THE WORLD!

JUAN SCHLIEPER

Black, brown, and Native Americans have never been treated fairly in America/the United States. No matter what people say. First it was slavery, then segregation. All and all, there has always been racism in the world. Not just in the United States, but all over the world. My question is, how do you think STEM could help with this age-old problem?

I believe that math could help us make calculations to prove that black and brown people are just as good as white people at many different things! I also believe that black and brown people can be amazing engineers if they want to be!

Science can also show us that there is nothing "worse" about black and brown people compared to white people!

And lastly, technology. Technology can help us advance like never before! This will lead to more jobs for everyone. More jobs mean a higher demand for workers, and that incentivizes employers to hire all the people they can find. This also means that if workers strike, it is devastating for the employer. Because of this, employers are incentivized to treat everyone fairly.

STEM CAN CHANGE THE WORLD BY

KAILOU BUCKHALTON

Evolving technology to support humanity. For example, a new way of doing things like work, transportation, and currency because every century evolves to create something new and innovative.

MICAH ALLEN

Science can be used to cure diseases. Technology can help people learn new information. Engineering can help build stores and houses. Math can help us understand finances. STEM continues to be a big part of our lives. In the future, STEM will grow to change every aspect for the better.

CAREERS THAT MAKE A DIFFERENCE

"We have the opportunity to create the future and decide what that's like."

Mae Jemison, former NASA astronaut

KYLAN JOHNSON

When I grow up, I think I'm going to be a good engineer. I am going to make robots, cars, TVs, and houses. I am going to make schools, stores, fire stations, police departments, hospitals, pet stores, and food places. I can make phones and tablets.

TERRAN CURTIS

STEM can change the world by many ways with technology and mathematics. Or engineering, because with engineering you can make buildings. With mathematics and technology you make a lot of new buildings and robots. With science, you can experiment.

STEM CAN CHANGE THE WORLD BY

MA'LAYAH LANE

Not being mean, not littering, being a teacher so you can teach the kids, being a judge so you can make things right, police officer so you can fight crime. You can be a science teacher. You can be a technology teacher. You can be an engineer who fixes cars, bikes, robots, and phones. You can teach kids mathematics just like you are a teacher.

STEM CAN CHANGE THE WORLD BY

MARION BUCKHALTON

Helping the world move into the future. It will help people in their jobs by making it easier and more productive.

It will organize teams. They will understand each other better and help each other do their jobs.

Teams make the job complete and achieve their collective goals.

STEM CAN CHANGE THE WORLD BY

DOMINIC WARE

Showing people not to litter and have them know why. Trying to make new types of medicine to heal people who do not feel good.

Helping people by making new inventions to solve people's body problems.

Using math to figure out another way to decompose trash so we have a better environment.

XAVIER BROWN

STEM can help the world.

How? Because it can help with young minds.

They can learn more and help with world problems like global warming and transportation like flying cars.

STEM: A CALL TO LEADERSHIP

"STEM education is very important to mankind just like the water we drink and the air we breathe in, we need it to produce more critical thinkers, problem solvers, innovators, and the world's best economy drivers."

Bamigboye Olurotimi, Technologist

STEM CAN CHANGE THE WORLD BY

NAJMA ABDI

Improving healthcare, reducing climate change, and all the problems in the world.

Learning about STEM can allow us to empower each other (especially young kids) to ask questions such as, "Why or how does something work the way it does?"

STEM can drive innovation and solve complex problems in various industries (for example: agriculture and healthcare).

I believe that teaching young kids about STEM will open doors to many opportunities and possibilities to making the Earth a better place.

STEM is the future!!

CHRISTOPHER JONES

We can change the world by not littering, leaving chemicals around, and all of the important things to keep the ground and Earth clean. So our air and surroundings are clean.

Everyone can have a part in this process to keep us safe.

WE CAN CHANGE THE WORLD BY

DA'RON PORTER

Finding cures for cancer.

Saving animals that are in danger.

Helping people who do not have homes.

Not leaving trash around in the sea because animals could choke or get strangled.

Keeping the world clean.

STEM CAN CHANGE THE WORLD

JAMAL WILLIAMS

Creating efficient avenues in life, taking past innovations, and making them more accessible to everyone. STEM is the grapeseed for innovation. By providing knowledge to the youth, we can water those seeds into grapes, and pass each person a grape to grow into their own grapeseed. Spreading knowledge to youth about the potential of STEM can help catalyze both old and new paying careers, which in turn can fuel our society.

REFLECTION GUIDE

These writing prompts can serve as a guide for exploring STEM education. Gather a group of students, friends, and family members to pause, reflect, and grow.

REFLECTION QUESTIONS

How can STEM change the world?

How can STEM promote justice and equity?

What are ways that STEM impacts your daily life?

Who are your STEM heroes and sheroes? How can they inspire your career journey?

How can you use STEM skills to build a better future?

ABOUT THE ARTIST

ELLA, A MINNESOTA GIRL WITH BIG DREAMS

Ella started Vibe Art (Visual Inspiration by Ella), as a means of sharing her love of art with others. For the last two years, she's been painting and creating fresh, energetic, colorful, beautiful art. Ella has now turned her artisanal talents into a true business, showing tweens that business ownership isn't just for adults.

Made in United States
Orlando, FL
05 November 2026

90230150R00019